The Colour
of Wine

For Kimberley

The Colour of Wine

Kevin Judd

CRAIG
POTTON
PUBLISHING

Photographs Kevin Judd

Introduction Michael Cooper

Seasonal introductions Kevin Judd

Publishing coordinator Robbie Burton

Design concept and cover Jo Williams

Filmwork Astra Print Ltd

Printing Everbest Printing, Hong Kong

First published in 1999 by Craig Potton Publishing,
PO Box 555, Nelson, New Zealand
Reprinted in 2001,2003

CONTENTS

THE VINEYARDS OF MARLBOROUGH

1. Allan Scott
2. Bladen Vineyard
3. Cairnbrae Wines
4. Cellier Le Brun
5. Clifford Bay Estate
6. Cloudy Bay
7. de Gyffarde Wines
8. Domaine Georges Michel
9. Drylands Estate Winery
10. Fairhall Downs Estate Wines
11. Forrest Estate
12. Foxes Island Wines
13. Framingham Wine Company
14. Fromm Winery
15. Gillan Estate

16. Grove Mill Wine Company
17. Hawkesbridge Wines
18. Herzog Winery and Restuarant
19. Highfield Estate
20. Huia
21. Hunter's Wines
22. Isabel Estate Vineyard
23. Jackson Estate
24. Johanneshof Cellars
25. Lake Chalice Wines
26. Lawson's Dry Hills
27. Le Brun Family Estate
28. Lynskeys Wairau Peaks
29. Montana Brancott Winery
30. Mount Riley

31. Mud House Wine Company
32. Nautilus Estate
33. Omaka Springs Estate
34. Ponder Estate
35. Saint Clair Estate Wines
36. Seresin Estate
37. Shingle Peak
38. Staete Landt
39. Te Whare Ra
40. Thainstone
41. Vavasour
42. Villa Maria Estate
43. Wairau River Wines
44. Whitehaven Wine Company

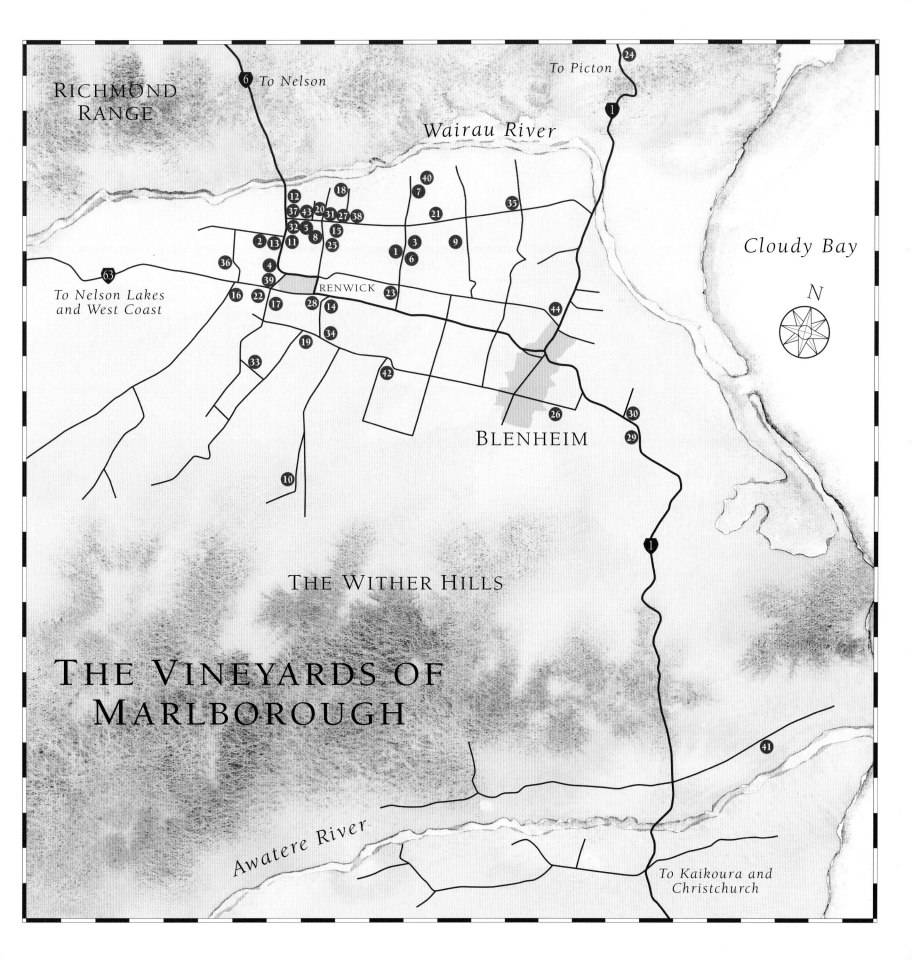

RICHMOND RANGE

6 To Nelson

To Picton

24

1

Wairau River

Cloudy Bay

12
18
37 43 20 31 27 38
40
7
35
32 5 8
21
2 13 11 15
25
3
1 9
6
36
4
N
39
RENWICK
23
16 22 17 28 14
44
34
19
26
30
33
BLENHEIM
29
42

10

1

THE WITHER HILLS

63

To Nelson Lakes
and West Coast

THE VINEYARDS OF
MARLBOROUGH

41

Awatere River

To Kaikoura and
Christchurch

Spring, Montana Brancott Estate

Summer

Autumn

Winter

VINEYARDS OF MARLBOROUGH

Michael Cooper

The pencil-thin plane noses across Cook Strait and the glistening Marlborough Sounds to where the Wairau River pumps into the sea at Cloudy Bay. Ahead stretches the pebbly, pancake-flat Wairau Valley, with countless green rows of vines softening the valley floor. Welcome to the home of the world's most striking recently discovered wine style – Marlborough Sauvignon Blanc.

In 26 exhilarating years since Montana began planting vines, Marlborough has emerged as New Zealand's most important wine region, with 36 per cent of the national vineyard. Vineyards are spreading like wildfire – from 1175 hectares in 1982 to 2071 hectares in 1992 and 3323 hectares in 1998.

Between 1996 and 2001, Marlborough's producing vineyards will expand by over 50 per cent.

Marlborough is New Zealand's – some say the world's – Sauvignon Blanc capital. Over two-thirds of all the country's plantings are in Marlborough, where Sauvignon Blanc is by far the most widely planted grape. With their leap-out-of-the-glass aromas and explosion of gooseberry and capsicum-like, mouth-wateringly crisp flavour, the Sauvignon Blancs here are of breathtaking intensity. According to British wine writer Hugh Johnson, "No region on earth can match the pungency of its best Sauvignon Blanc."

Dog Point Vineyard, Brancott Valley

10

If you visit Marlborough, you can feel the optimism in the air. Export orders are flooding in and the quality of recent harvests has generally been high. The aristocracy of the local wine scene includes some of the biggest names in New Zealand wine – Montana, Corbans, Cloudy Bay, Hunter's. The community of Marlborough-based wine producers is growing swiftly: nine in 1990, 20 in 1993, 56 in 1999.

Overseas investment is streaming into the region. Australian winemakers have been active in Marlborough since 1985, when David Hohnen of Cape Mentelle set up Cloudy Bay to supply the Australian market with Marlborough Sauvignon Blanc. Corbans Marlborough Winery, now wholly Corbans-owned, was originally funded by Corbans and Wolf Blass, and Domaine Chandon of Victoria produces a brilliant bottle-fermented Marlborough Brut.

European winemakers are also scattered around the region. Almuth Lorenz, raised in the Rheinhessen, is owner of the Merlen label; Edel Everling, co-founder of Johanneshof Cellars, was born at Rudesheim. The Fromm winery, specialising in red wines, was set up by Swiss immigrants, Georg and Ruth Fromm.

Most conspicuous are the mounting links between Marlborough and the great houses of Champagne. Deutz Marlborough Cuvee is produced by Montana under the technical guidance of Deutz and Geldermann. Veuve Clicquot is now the majority shareholder in Cloudy Bay. Moet & Chandon, owner of Domaine Chandon, is involved in masterminding the production of Domaine Chandon Marlborough Brut. Sauvignon Blanc is currently the star of the Marlborough wine scene, but long-term, the region's greatest gifts to the wine world may well include its classy bottle-fermented sparklings.

For a wine region with such a big reputation, Marlborough (or at least the Wairau Valley, where most of the vines are concentrated), is remarkably small. Rugged mountains up to almost 3000 metres straddle most of Marlborough. The Wairau Valley, 14 kilometres wide at its eastern extremity, where it meets the sea at Cloudy Bay (so named because after heavy rain, its waters turn cloudy with silt from the Wairau River), runs 26 kilometres inland. From the Wither Hills to the south to the towering Richmond Ranges on the valley's northern flanks is only about ten kilometres.

When Montana planted its first vines in the province, it triggered the modern era of Marlborough viticulture. The region's first wines, however, had flowed almost a century earlier. David Herd's Auntsfield vineyard, in the hills to the south of Fairhall and Brancott, produced its first commercial harvest around 1875. Auntsfield's sweet red wine was made from red Muscatel grapes,

"crushed with a machine made from the wheels of an old flax mill stripper, then pressed in a barrel... [and] matured in oak brandy casks." (Cynthia Brooks, *Marlborough Wines and Vines*, 1991.) Only about 800 litres were produced each vintage, but the trickle of Auntsfield wine survived Herd's death in 1905; his son-in-law Bill Paynter carried on the family tradition until 1931.

At Mount Pleasant Wine Vaults, just south of Picton, in 1880 George Freeth started making wine from a wide array of fruits, including grapes. Yet no surge of vine plantings in Marlborough followed the 1895 publication of Romeo Bragato's influential *Report on the Prospects of Viticulture in New Zealand*; he was more impressed with Nelson's potential. In the first half of the 1900s, in the heart of Blenheim, Harry Patchett and Mansoor Peters grew grapes and sold a trickle of wine. Patchett lived until 1974 – just long enough to witness the Montana-led revival of Marlborough wine.

"In 1973, Marlborough was heavily reliant on sheep farming," recalls Philip Gregan, chief executive officer of the Wine Institute of New Zealand. "The name 'Marlborough' was practically unknown outside New Zealand, while within New Zealand it was best known for rugby players... Into this quiet, sunny, rural environment in 1973 came Frank Yukich, Wayne Thomas and Montana Wines."

Montana's pivotal move into Marlborough followed its whirlwind expansion in the 1960s, during which it planted the country's biggest vineyard at Mangatangi, in the Waikato, and arranged major grapegrowing contracts with Gisborne farmers. As wine sales boomed, Montana's grape needs intensified. Wayne Thomas, then a scientist in the Plant Diseases Division of the DSIR, has related: "Although plenty of suitable land was available in both the Poverty Bay and Hawke's Bay regions, my own impression was that it was too highly priced for vineyards." Thomas suggested the Marlborough region as an alternative, and soon after the area's suitability for growing wine grapes was independently confirmed by the Viticulture Department at the University of California, Davis.

Wayne Thomas's 1973 report on the region's viticultural potential was a decisive factor in Montana's move into Marlborough, but earlier others had speculated about the possibilities of Marlborough wine. "One undeveloped area with distinct possibilities for viticulture is behind Blenheim, among the northern foothills of the Kaikouras," John Buck wrote in his book *Take a Little Wine*, in 1969. Laurie Millener concluded in 1972, after analysing New Zealand's regional climates, that "Nelson, and especially Blenheim, offer great promise. It should be possible to make specialist table wines there, both reds and 'Mosels.'"

Marlborough was often cited by politicians of that era as a typical example of a 'forgotten' region needing diversification. "Nothing more vividly recalls the sudden realisation of what wine could do for Marlborough," Terry Dunleavy, Montana's sales manager in the early 1970s, has written, "than the stunned reaction of Lucas Bros. when faced with a [Montana] order for twenty-six tractors."

The first vine was planted in Marlborough on 24 August 1973: a silver coin, the traditional token of good fortune, was dropped in the hole and Sir David Beattie, then chairman of the company, with a sprinkling of sparkling wine dedicated the historic vine. The first grapes were harvested on 15 and 16 March 1976; 15 tonnes of Müller-Thurgau were trucked aboard the inter-island ferry at Picton and driven through the night by Mate Yukich (Frank's brother) to Montana's Gisborne winery. A 'token' picking of Cabernet Sauvignon followed in April.

In 1979, Montana bottled its first Marlborough Sauvignon Blanc; the first commercial release flowed in 1980. Overflowing with fresh-cut grass aromas and flavours, those pungent early vintages of Sauvignon Blanc were without parallel anywhere else in the wine world, even in the Loire (France's traditional Sauvignon Blanc stronghold); Marlborough was on its way to international stardom.

Penfolds (NZ), then one of the country's largest wineries, and Corbans established vineyards in Marlborough several years after Montana. Penfolds' first contract vineyards were planted in the winter of 1979 and subsequently the company arranged contracts amounting to about 400 hectares. Behind Penfolds' expansion was the hard-driving Frank Yukich, who had parted company with Montana in 1974. However, Penfolds' plans for a Blenheim winery were ultimately shelved and, until Montana purchased the firm in 1986, Penfolds' Marlborough grapes were trucked to the North Island for processing.

Corbans arrived in Marlborough in 1980, planting its own vineyards, notably Stoneleigh. Unlike Montana, which had established its vineyards on the south side of the Wairau Valley, where the less gravelly soil was thought to be kinder to machinery, Corbans planted its vines on the north side of the valley, in the stonier Rapaura district.

For Jane Hunter (today the most famous woman in the New Zealand wine industry), much has changed since January 1983, when she arrived in Marlborough to take up the job of Montana's chief viticulturist. "There were Penfolds growers, Montana growers, Corbans growers; then there were Te Whare Ra, Hunter's and Daniel Le Brun; and that was it. Next on the scene was Cloudy Bay. In the early days, it was mainly Müller-Thurgau, although Montana did have some Sauvignon Blanc at Brancott and Ernie [Hunter] had seven and a half acres of Sauvignon. It used to amuse me that after 1986, people would fly in here from overseas saying 'We want to see the Sauvignon Blanc,' and there was so little of it here. All the fuss was being made over a tiny few acres. Then came the grape pull [the government-funded vine uprooting programme of 1986], and a lot of other varieties were grafted over to Sauvignon and Chardonnay."

'The fuss' (as Jane Hunter puts it) about Marlborough Sauvignon Blanc in the mid 1980s was caused by an extroverted, entrepreneurial Irishman, Ernie Hunter, who founded Hunter's Wines in 1982. With Almuth Lorenz as winemaker, the fledgling winery made an eye-catching debut at the 1982 National Wine Competition by entering six wines and collecting six medals. However, beset by financial problems, in 1984 Ernie Hunter turned to export and shipped thousands of cases of wine to the UK, USA and Australia.

In his widow, Jane's, words: "Ernie didn't just sell Hunter's wines. When he was in New Zealand he always talked about Marlborough wines and when he was overseas he talked about New Zealand wines." Hunter's most publicised successes came at the 1986 and 1987 Sunday Times Wine Club Festivals in London, when the public voted his 1985 Fumé Blanc and 1986 Chardonnay as the most popular wines of the show.

A man of formidable energy and charisma, and a pioneer of New Zealand wine exports, Ernie Hunter died in a motor accident in 1987, aged 38. Jane stepped in as the managing director, and today Hunter's ranks among the country's most highly respected medium-sized wineries.

The key ambassador for Marlborough wine in world markets has been Cloudy Bay. When David Hohnen moved to establish a new winery in Marlborough, the New Zealand wine industry was on the verge of a ferocious, glut-induced price war. "It was a terrific gamble," he recalls. "I just had this gut feeling that told me it was the right thing to do. New Zealand Sauvignon Blanc simply hadn't been discovered (overseas) and seemed to me to have a great future." Explosively flavoured and stunningly packaged, Cloudy Bay's first 1985 Sauvignon Blanc swiftly sent a ripple through the international wine world.

Over the years, Cloudy Bay Sauvignon Blanc has been hailed as "New Zealand's finest export since Sir Richard Hadlee" (David Thomas in *Punch*), and as "like hearing Glenn Gould playing the Goldberg variations or seeing Niki Lauda at full tilt" (Mark Shield in the *Sun Herald*, Melbourne.) The annual output of Cloudy Bay Sauvignon Blanc, New Zealand's most internationally prestigious wine, now exceeds 40,000 cases.

An even greater number of overseas wine lovers have been introduced to the delights of New Zealand wine by Montana Marlborough Sauvignon Blanc, which places its accent squarely on the explosive flavours of slowly ripened Marlborough grapes. Grown in the sweeping Brancott vineyard, machine-harvested and cool-fermented in stainless steel tanks, this is a zesty, assertive style of Sauvignon Blanc, packed with green capsicum-like flavour, appetisingly crisp and fresh. The 1989 vintage won the Marquis de Goulaine Trophy for the champion Sauvignon Blanc at the 1990 International Wine and Spirit Competition in London; the 1991 vintage was crowned the champion white wine of the 1992 Sydney International Winemakers' Competition. The output of Montana Marlborough Sauvignon Blanc ranges from 120,000 to 150,000 cases per year, of which over 100,000 cases (1.2 million bottles) are exported.

Corbans each year produces over 85,000 cases of its acclaimed Stoneleigh Vineyard Marlborough Sauvignon Blanc, of which about half is exported. The company reports that Stoneleigh Vineyard Sauvignon Blanc is on sale in more corners of the globe than any other New Zealand wine.

Oz Clarke, a British wine author and television personality, conveyed in his usual colourful style the strong impact of Marlborough Sauvignon Blanc in the UK, during a Radio New Zealand interview in the early 1990s. "The great thing about Marlborough is that it produced for the first time since the War, maybe this century, a flavour which no-one's ever found before. Marlborough flavour is unbelievably strong, unbelievably memorable... The Sauvignon is absolutely, stabbingly strong fruit: it's a mixture of apricots and asparagus and grassiness which is terribly exciting because the Sauvignon grape in Europe, which they've been growing for donkey's years, has got drier and drier and leaner and leaner... Suddenly, out comes Marlborough with this flavour which is so strong."

What is the magic ingredient that enabled Marlborough's winemakers to take the world by storm? A temperate climate in which the grapes ripen slowly but fully, conjuring up their most intense aromas and flavours, is Marlborough's key viticultural asset. Blenheim frequently records the highest total sunshine hours in the country. To counter the dehydrating effects on the vines of hot, dry north-westerly winds, grapegrowers in the Wairau Valley draw irrigation water from an extensive aquifer within the valley's alluvial gravels. The combination of warm, sunny days and clear, cold southern nights means that even when the grapes' sugar levels are rising swiftly, their acid levels stay high, giving the wines their refreshing crispness and zing.

Marlborough's autumn weather is typically drier than in North Island regions and during the critical harvest month of April, the average temperature is quite low. This slows the spread of disease and the grapes can usually be left on the vines for an extended ripening period.

Barriques

Yet Marlborough's 200-odd grapegrowers and 56 wine companies (in mid 1999) often grapple with weather extremes. A heavy autumn frost in 1990 killed the vines' leaves, preventing the full ripening of Riesling, Cabernet Sauvignon and Sauvignon Blanc: "The whole valley turned black overnight," recalls Jane Hunter. The summer of 1993 was abnormally cold. In 1995, incessant rains in April played havoc with the harvest. So exceptionally drought-stricken was the summer of 1998, camels were reportedly spotted on the horizon.

Not all the soils found in Marlborough are well-suited to viticulture. The Wairau Valley is a flood plain, with braids of soil of varying fertility. Soil types often vary enormously even within individual vineyards. Free-draining, shingly sites are the most sought after. "The key benefit of the stones is that they reduce the soil's fertility," says John Belsham of Foxes Island.

Marlborough's original grape plantings were on the southern fringes of the Wairau Plains, but in 1977 Phil and Chris Rose (who now own the Wairau River wine company) pioneered grapegrowing on the north side of the valley. After attracting 56 objections from local farmers (many of whom are now grapegrowers) concerned that their own hormone sprays would damage vineyards, the Roses took the issue to the Court of Appeal and "basically won hands down."

Vavasour in the mid-late 1980s pioneered viticulture in the

Awatere Valley, east and over the Wither Hills from the more sweeping Wairau Valley. According to viticultural consultant Richard Bowling, "climatically the Awatere Valley is akin to the Wairau, the Dashwood having a tendency to be drier than Rapaura. The braided river pattern resulting in very uneven [soil] profiles in the Wairau Valley is also much less pronounced..." However, a water supply problem – there is no aquifer, only a seasonal, snow-fed river, which slows to a trickle in a drought year – has slowed the expansion of vineyards in the Awatere.

Vineyards in the Wairau Valley have also spread onto land with a high water table east of Blenheim (which being built on a swamp, with the first houses on stilts, was originally called Beaver Town.) Further west, the lower Waihopai Valley, where the frost risk is greater but land is cheaper, is currently attracting a flurry of vine planting. Brian Bicknell, of Seresin Estate, sees plenty of scope for further vineyard spread. "The hills have not been touched. As you get higher there is more clay and a problem getting water, but the slopes offer the potential of a good aspect."

There is a growing awareness that exactly where in Marlborough vines are planted can have a significant effect on wine styles, reflecting local variations in soil and climate. In the stony soils on the north side of the Wairau Valley, Sauvignon Blanc grapes ripen earlier than those on the less gravelly south side. "If you want to make a very herbaceous, greener style of Sauvignon Blanc," says John Belsham, "you choose a site with medium-high fertility and good soil moisture retention. For a lusher, less aggressive Sauvignon Blanc, you choose a stony site which matures its fruit two to three weeks earlier."

To call Marlborough Sauvignon Blanc country is an oversimplification. Over two-thirds of New Zealand's Sauvignon Blanc is planted in the region, and Sauvignon Blanc and Chardonnay together account for 67 per cent of all Marlborough's vines. But the province's 3323 hectares of vines in 1998 also included substantial areas of (in order): Pinot Noir, Riesling, Merlot, Semillon, Cabernet Sauvignon, Müller-Thurgau, Gewürztraminer and Pinot Gris.

Sauvignon Blanc dominates (its 1382 hectares in 1998 accounted for nearly 42 per cent of the region's vines), but Marlborough's plantings of the prestigious Chardonnay, Riesling and Pinot Noir varieties are heavier than in any other part of New Zealand. Over half the country's Riesling is concentrated in Marlborough; so is 46 per cent of the Pinot Noir. The areas devoted to Müller-Thurgau (which in 1986 was Marlborough's most widely planted grape, but now forms only three per cent of the bearing vineyard), Semillon

and Cabernet Sauvignon are all contracting, but Gewürztraminer and Pinot Gris are both on the rise.

Viticulture is flourishing in Marlborough, but it hasn't been all plain sailing. The region's vineyards have recently been severely affected by phylloxera, a parasitic disease of the vine that rampaged like a prohibitionist zealot through French vineyards in the 1870s. The aphid, first identified in New Zealand in 1895, wiped out many of the country's early vines by attacking their root systems. In January 1984, phylloxera was detected on the roots of three weak vines in Marlborough.

Phylloxera spread swiftly through the Wairau Valley, reducing the vines' yields and retarding ripening. The only effective way to combat phylloxera is to graft vines onto phylloxera-resistant rootstocks, but in 1994 just 27 per cent of Marlborough's producing vines were grafted (ungrafted vines are a lot cheaper.) "Phylloxera has been very damaging in Marlborough," believes Ivan Sutherland, viticulturist for Cloudy Bay and a long-term grapegrower in his own right. "Replanting is very costly and it has a devastating effect on production levels." However, following massive replanting, in 1998 virtually 90 per cent of Marlborough's vines were grafted.

Although the key to Marlborough's wine quality lies in the vineyards, the grapes' fresh, incisive flavours and fine-wine potential must still be preserved through the various stages of winemaking. "It was fortunate that Marlborough emerged at a time when reductive [non-oxidative] winemaking techniques, first taught at the University of California, Davis, were being adopted as the New Zealand and Australian style of winemaking," says Brian Bicknell of Seresin Estate. "By using these reductive techniques, it is relatively easy to ensure that all the fruit characters of the juice are carried through to the bottle. These fruit-driven wines caught the public's imagination... [because] all of a sudden it was easy to see characters in wines that everyone could associate with, and you did not need years of experience to start connecting these wines with varieties and regions."

Marlborough's fast-multiplying wine companies vary in size from tiny to leviathan. Some of the smallest are 'grower labels'– a term used when a former specialist grape-grower keeps back part or all of the crop, has the wine made on his or her behalf at a local winery, and then controls the marketing of the brand. These producers often only make a few hundred or a few thousand cases of wine per year.

At the opposite end of the scale is Montana's winery at Riverlands, a few kilometres on the seaward side of Blenheim. The first winery in the region, it is still by far the largest. Montana's 'tank farm', with towering 550,000-litre insulated tanks, has the capacity to store up to 20 million litres of Marlborough wine.

Until 1995, the Vintech contract winemaking facility handled about one-third of the entire Marlborough harvest – simply de-juicing grapes for some of its 30 clients, but fermenting, processing and bottling wine for others. After Vintech was purchased by four producers – Matua Valley, Nautilus, Foxes Island and Wairau River – the new company, Rapaura Vintners, halved its list of clients. Some of those who were forced to go elsewhere fast-tracked their development plans and built their own wineries.

Apart from the many wineries with a physical presence in the region, a host of companies based in other regions, from Auckland to Central Otago, also include Marlborough wines in their range. Goldwater Estate, on Auckland's Waiheke Island, which in most years produces 500 to 1,000 cases of its classic Cabernet Sauvignon/Merlot, in 1999 made over 17,000 cases of its Dog Point Marlborough Sauvignon Blanc. "Our agent in England asked us for some Marlborough white wines to supplement our red," says Jeanette Goldwater, who reports that the Sauvignon Blanc is now also enjoying strong demand in the US.

Sauvignon Blanc has been the glittering success, but there is more than one string to Marlborough's vinous bow: the region also produces many of New Zealand's classiest Chardonnays, Rieslings, botrytised sweet whites and bottle-fermented sparkling wines.

Pinot Meunier

Marlborough's Chardonnays are fresh, vibrantly fruity and threaded with appetising acidity. The Rieslings are strongly scented and mouth-wateringly crisp, with intense lemon/lime flavours.

With their piercing flavours and tense acidity, Marlborough grapes are ideally suited to the production of top-class sparkling wines. The region's finest bubblies typically offer the rich fruit flavours that are pure Marlborough, coupled with impressive delicacy, lightness and finesse. Marlborough is also the source of many of New Zealand's most ravishingly beautiful honey-sweet whites, which have been shrivelled and dehydrated on the vines by 'noble rot', the dry form of the *botrytis cinerea* mould.

Marlborough's versatility as a wine region has recently been underlined with several highly promising Pinot Gris and Gewürztraminers. Cabernet Sauvignon's performance has been disappointing, reflecting the fact that you can't grow explosively flavoured, herbaceous Sauvignon Blanc and rich, fully ripe Cabernet Sauvignon side-by-side. Pinot Noir and Merlot, which ripen significantly earlier than Cabernet Sauvignon, hold the keys to Marlborough's red-wine future.

The signs of its wine success are everywhere in Marlborough: in the leafy-green landscape; in the striking architecture of many of the wineries and their restaurants; and most importantly, in the wines themselves. Wine is big business, according to a 1997 study by Lincoln University, which found that Marlborough's grape and wine industry generated direct sales of $130 million per year, of which $70 million was returned to the region by way of purchases of goods and services. Each year, over 250,000 tourists flock to the region's wineries.

Where will it all end? Noel Scanlan, chief executive of Corbans, has estimated that half of Marlborough's potential viticultural land has already been planted and that, at the current rate of expansion, Marlborough will reach its full viticultural capacity around 2013. In terms of size, the region will never be another Barossa Valley or Bordeaux.

Marlborough's future lies as a small-scale producer of distinctive, flavour-crammed wines, full of personality. The brand 'Marlborough Sauvignon Blanc' is already a force in world markets.

It's still early days for Marlborough wine. A lot of learning lies ahead, both in viticulture and winemaking. New grape varieties and new sub-regions are waiting to be explored. The quality of the wine can only get better as the vines age, anchoring their roots deeper and deeper in Marlborough soil.

Spring

Tiny buds begin to swell as sap starts to bleed from the last of the season's pruning cuts. Woolly and fragile the first leaves emerge and the cycle of the vineyard begins again.

Row upon row, the vines slowly awaken from their dormant winter state. In the low evening light the young shoots create their own luminance, transforming the previously lifeless landscape into a glittering gold and lime patchwork. Vibrantly green, the young shoot tips elongate, tiny inflorescences unfold and the first signs of the new year's crop become visible. Spring rains and gradually rising temperatures accelerate the flurry of growth and before long the vines' outstretched shoots are being tossed around by Marlborough's infamous nor'west winds. Whilst the landscape seems to flow with the gusts, the vineyard teams struggle against it, lifting foliage wires to contain the new growth.

Flower caps eventually shed and the pistils are exposed to the elements - flowering has commenced and conception begins.

ABOVE: Chenin Blanc
LEFT: Spring, Montana Brancott Estate

ABOVE: Budburst, Cabernet Sauvignon
RIGHT: Young leaves after spring rain, Cabernet Sauvignon

ABOVE: Spring growth, Brancott Valley
RIGHT: Nautilus Estate, Awatere River Vineyard

Tips of shoots, Chenin Blanc

Spring sunset, Stoneview Vineyard, Rapaura

ABOVE: Young vines, Medway River Vineyards, Awatere Valley
LEFT: Stoneleigh Vineyard, Rapaura

Wildflowers, Brancott Valley

Northwesterly, Dog Point Vineyard, Brancott Valley

Last light, Stoneleigh Moorlands Vineyard, Rapaura

Unfolding leaf, Sauvignon Blanc

Summer

Bright blue skies, clear sunny days. The stony, alluvial soils slowly dehydrate and the surrounding hills lose their lushness, taking on a golden-straw hue that contrasts starkly with the carpet of dark green vineyard below. Contained within their trellis, the vines are now trimmed by an army of tractors, transforming the scruffy uncontrolled vegetation into manicured hedgerows that roll across the contours of the Wairau Valley.

Steadily, the grape berries enlarge until they reach véraison, the point where ripening commences and red varieties lose their iridescent greenness, taking on a deep, plum-like colour.

ABOVE: Sauvignon Blanc
LEFT: Summer, Montana Brancott Estate

From this time on sugar begins to accumulate in the fruit. The tranquillity of the vineyard is lost as the battle against the elements begins. An amazing array of scarecrows suddenly appears to stand guard over the crop, and gas-fired cannons boom, reverberating around the valleys. Shot gun-carrying motorcyclists roar up and down the rows and teams of workers move through the vineyards plucking leaves to expose the fruit to the full intensity of the sunlight.

As ripe corn is being cut and strong aromas waft from the valley's garlic sheds, the winemakers watch the sky. They prowl among the vines, tasting and sampling as ripening progresses, until eventually the moment is right, and the vintage begins.

ABOVE: Lower Brancott Valley, looking towards Mt Riley
RIGHT: Lenticular clouds, Rapaura

Long white cloud, Brancott Valley

Summer sky, Kegworth Vineyard, Woodbourne

Sauvignon Blanc cluster after summer rain

Pinot Noir during véraison

The Vavasour Vineyard, Awatere Valley

Richmond Ranges and apricot orchard, Rapaura

ABOVE: Summer burn off, Richmond Ranges
LEFT: Roadside wildflowers, Motukawa Farm, Rapaura

Medway River Vineyards, Awatere Valley

Brackenfield Estate Vineyard, Awatere Valley

Vineyards at dawn, lower Awatere Valley

Stooked oats, Willowhaugh Farm, Woodbourne

Scarecrow and harvester, Riverbrook Vineyard, Rapaura

Scarecrow family, Hawkesbury Valley

'Mother crow'

Autumn

The harvest commences quietly. Brightly coloured picking bins line vineyard rows and teams of pickers follow their trail, filling them with the ripe clusters. Gangly machine harvesters emerge from their sheds to begin the annual marathon, straddling the seemingly endless crop-laden rows.

As the nights become cooler the pace begins to hasten. Convoys of tractors and trucks roam the valleys at all hours of the day, collecting the precious crop. The harvesters rattle and shake their way through the night, gathering grapes for wineries lit up like fairgrounds, that hum with the sound of machinery and heavy metal. Intoxicating vinous aromas drift through the still of the evening, as the fresh grape juice is transformed into wine.

Soon the green vineyard landscape yellows as the vines are relieved of their crop. Days grow shorter and the sun shines lower in the sky. The colours intensify and the golden landscape basks in the warmth of the clear autumn light.

ABOVE: Cabernet Sauvignon
LEFT: Autumn, Montana Brancott Estate

Autumnal patterns, Brancott Estate

Autumnal patterns, Brancott Estate

Widelah Vineyard, Brancott Valley and the Wither Hills

Clayvin Vineyard, Brancott Valley

ABOVE: Red grape details
LEFT: Malbec

ABOVE: Shingle Peak Cottage Vineyard, Rapaura
RIGHT: Cabernet Sauvignon

Greywacke Vineyard, Rapaura

Cob Cottage ruins, Squire Estate Vineyard, Rapaura

ABOVE: White grape details
RIGHT: Sauvignon Blanc

ABOVE: Scarecrow, Clayvin Vineyard, Brancott Valley
LEFT: Hot air balloon, Brancott Valley

Cabernet Sauvignon

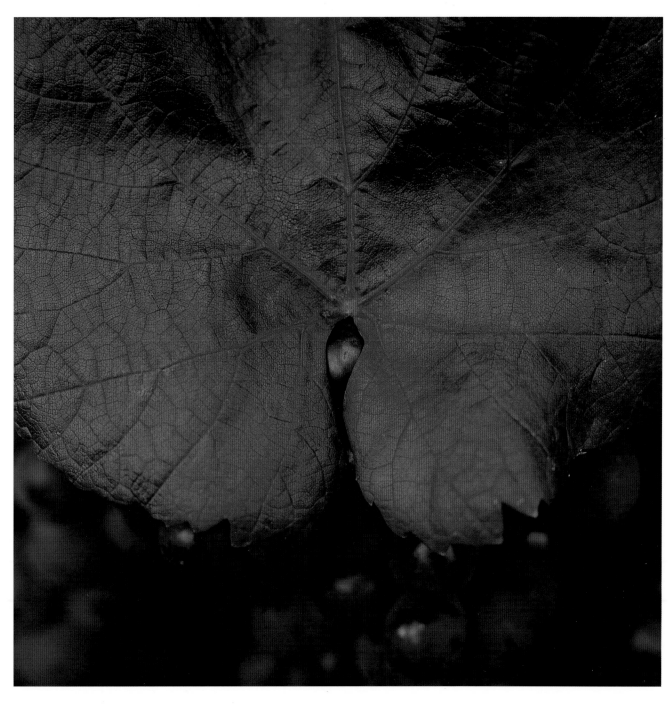

Pinot Noir

Simmerland Vineyard, Rapaura

Sheep grazing on the foothills, Brancott Valley

ABOVE: Shiraz
LEFT: Montana Renwick Estate Vineyard

ABOVE: Cloudy Bay Vineyard, Rapaura
LEFT: Merlot

ABOVE: Hand-picked bunches of ripe grapes
RIGHT: Cabernet Sauvignon

Second set, Pinot Noir

Ripe second set, Pinot Noir

Allan Scott Winery, Rapaura

Sticky fingers

Machine harvesting, Brancott Estate

Brancott Estate

ABOVE: Babich Wakefield Downs Vineyard, Awatere Valley
RIGHT: Nautilus Estate Awatere River Vineyard and Mt Tapuaenuku

Winter

As the snow returns to the distant peaks and the last of the autumn leaves fall, the only fruit remaining on the vines is that which has been left to 'rot'. It has not been forgotten however. In some years a fungus, 'noble rot' (*botrytis cinerea*), covers the berries in a furry coat that saps and shrivels the grapes, concentrating the juice into sticky nectar. Carefully hand harvested, this raisined fruit will yield little juice, but produces lusciously sweet dessert wine.

With the arrival of the first frosts the vineyard once again becomes a hive of activity. White and glistening, the frosty valley floor contrasts with the naked, bronzed canes that stand glowing in the warm light of dawn. Wrapped in many layers of clothing, the pruning gangs trudge their way down the rows, wielding sharpened secateurs to make their calculated cuts.

Then come the 'strippers', an energetic bunch that work their way through the vineyard wrenching out the unwanted canes that cling so securely to the trellis with their wiry tendrils. Finally, the agile 'tying down' teams wrestle with the skeletal remains of the pruned and stripped vines, securing them gently but firmly to the trellis wires, where they await next season's rising of the sap.

ABOVE: Frost on Chardonnay leaves
LEFT: Winter, Montana Brancott Estate

Dawn frost, Brancott Valley

Glowing canes, Brancott Valley

Riesling

Riesling with *botrytis cinerea*

ABOVE: Winter moon, Brancott Valley
RIGHT: Frosted leaves

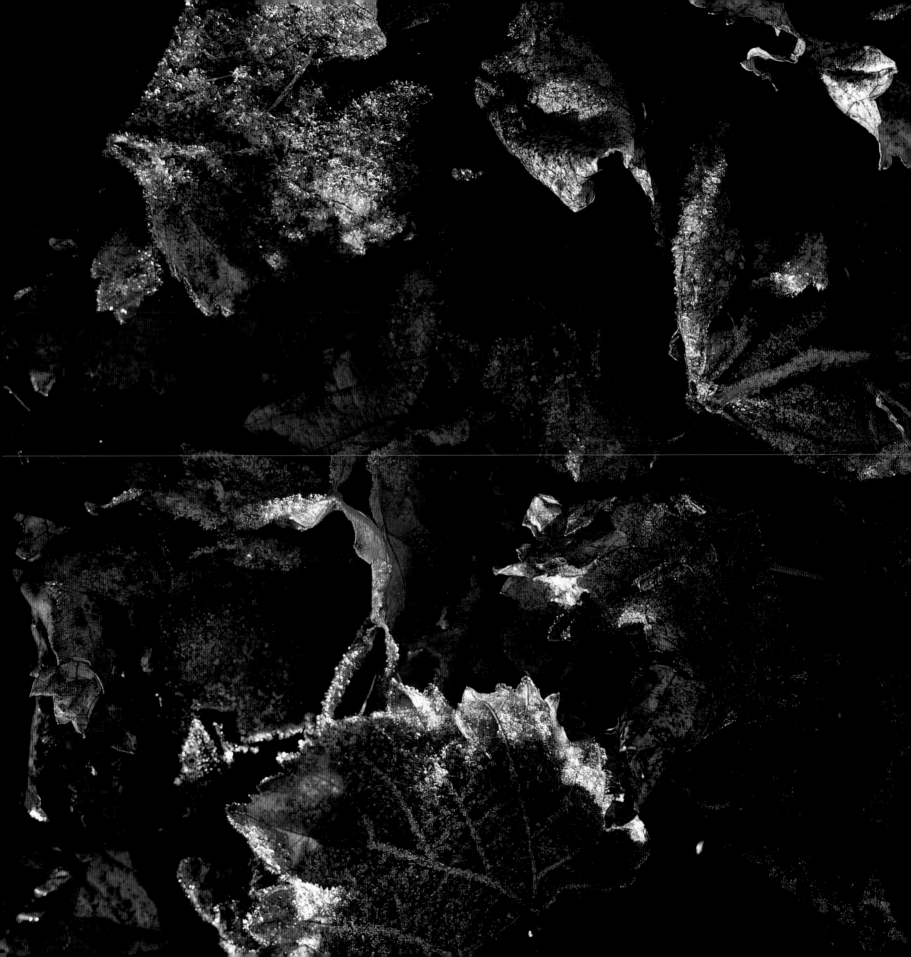

ABOVE: Mist, Clayvin Vineyard, Brancott Valley
RIGHT: Pruned vines, Stoneleigh Moorlands Vineyard, Rapaura
PREVIOUS PAGES: Pruning, Motukawa Vineyard, Rapaura

ABOVE: *Botrytis cinerea* on Riesling
PREVIOUS PAGES: 'Wine Thief', Cloudy Bay Winery (Photo: Kohen Judd)

Cobweb in Riesling vineyard

Dawn frost, Brancott Estate

Dawn frost, Brancott Estate

ABOVE: Pinot Noir leaves
RIGHT: Corkscrew tendril

ABOVE: Bonovoree Vineyard, Blenheim
RIGHT: Barriques

Fairhall Downs Estate vineyard, Brancott Valley

Sunset over the Richmond Range

Chardonnay

ACKNOWLEDGEMENTS

I would like to express my gratitude to several special people for helping me to achieve something that I had never dreamed I would do:

To my father Ron, thanks for exposing me to photography at an early age and my mother Freddy for my development. To Kimberley for your all-round support of my obsession, and my two boys Kohen and Alex for your companionship and patience. To Mick Rock for inspiration and direction. To David Hohnen for allowing me to pursue a passion and Jane Adams for encouragement and assistance. To Julie Dalzell for first considering one of my images to be cover material. To Robbie Burton for his hard-line but slightly flexible picture selection talents. And lastly, to all the photographers whose brains I have picked over the years and all the vignerons of Marlborough for allowing me access (some unknowingly) to their vineyards.

Kevin Judd
Blenheim, September 1999

TECHNICAL DETAILS

The vast majority of the images were taken using Bronica 6x6 cm systems, initially with my old Bronicasaurus S2A and more recently an SQAi, with lenses ranging from 50mm to 250mm and extension tubes used for close ups. The rectangular images were shot with a Nikon FE and the panoramic photographs on pages 94/95 and 104/105 were taken using Widelux F6 and Linhof Technorama respectively. Almost all pictures were shot on Fujichrome Velvia transparency film. A polarising filter was used on a number of the landscape photographs, and the only other form of filtration was the occasional use of graduated neutral density to balance sky and foreground exposure.

KEVIN JUDD PHOTOGRAPHY

For further information or photographic print enquiries, please contact:
Kevin Judd, PO Box 1189, Blenheim,
Marlborough, New Zealand.
Email: kevinjudd@xtra.co.nz
Website: www.kevinjudd.co.nz

The Judd brothers